Tsunamis

BY CHRISTY STEELE

Steadwell Books

Raintree Steck-Vaughn Publishers
A Harcourt Company

Austin · New York
www.steck-vaughn.com

Published by Raintree Steck-Vaughn Publishers, an imprint of Steck-Vaughn Company.

Library of Congress Cataloging-in-Publication Data
Tsunamis/by Christy Steele.
    p.cm.—(Nature on the rampage)
Includes bibliographical references and index.
ISBN 0-7398-4706-6
1. Tsunamis—Juvenile literature. [1. Tsunamis.] I. Title. II. Series.
GC221.5 .S74 2001
551.47'024—dc21

2001019511

Printed and bound in the United States of America
1 2 3 4 5 6 7 8 9 10 WZ 05 04 03 02 01

Produced by Compass Books

Photo Acknowledgments
NOAA, cover, title page, 4, 8, 20, 23, 24, 26, 29
Photo Network, 10

Content Consultants
Dr. George Curtis   Maria Kent Rowell
Division of Natural Sciences   Science Consultant
University of Hawaii at Hilo   Sebastopol, California

David Larwa
National Science Education Consultant
Educational Training Services
Brighton, Michigan

This book supports the National Science Standards.

CONTENTS

A tsunami has smashed these buildings into broken pieces.

When the Water Rises

A tsunami is a group of very deep and very long waves. A tsunami's waves travel out in circles from the point where it started. The waves move along in a wave train, one wave following another.

A tsunami's waves are only 3 feet (0.9 m) high in the open ocean. The waves become very steep as they reach shore. Near land, the waves can rise up to 100 feet (31 m) or more.

Powerful tsunamis create powerful, killer waves. They wash away anything in their path. They can snap trees like toothpicks, crush houses, and toss large boats and rocks onto shore. A tsunami can flood large areas of low-lying land.

About Tsunamis

Tsunamis are huge columns of water that may stretch thousands of feet down into the ocean. They stretch all the way down to the ocean floor.

Tsunamis flow outward at great speeds. In the deep ocean, they move up to 500 miles (805 km) per hour. There may be hundreds of miles between each tsunami wave. The series of waves races across thousands of miles of open ocean until they reach land. They can travel from one end of the ocean to the opposite shore in less than one day.

Near the coast, where the ocean is not as deep, tsunamis begin to change. Waves bounce off the ocean floor. This makes the tsunami slow down as it nears the coast. The waves of a tsunami are pushed closer together. A tsunami grows taller as the ocean becomes shallow. The waves grow taller and taller until they finally wash over land.

The shape and height of tsunamis depends on the shape of the ocean floor and coastline

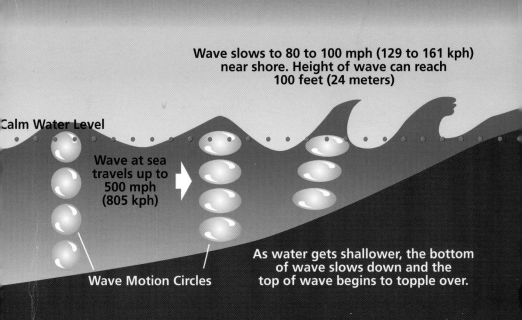

Wave slows to 80 to 100 mph (129 to 161 kph) near shore. Height of wave can reach 100 feet (24 meters)

Calm Water Level

Wave at sea travels up to 500 mph (805 kph)

Wave Motion Circles

As water gets shallower, the bottom of wave slows down and the top of wave begins to topple over.

This illustration shows how tsunamis behave as they approach land.

where they strike. In some places, offshore **reefs** may slightly break the force of a tsunami. A reef is a strip of sand, rocks, or coral that rises from the ocean floor almost to the surface of the water. In other places, the seafloor may contain a deep **valley**. A valley is a deep crack in the seafloor. Since a valley is deeper than other areas of the seafloor, the tsunami will not grow as tall there.

▲ A tsunami in Alaska has washed this boat onto the shore.

What a Tsunami Does

Each tsunami acts differently when it strikes. Some tsunamis draw water away from the shore right before they hit. People may see fish flopping on the sand. Other tsunamis wash over land with no warning.

Some people think that tsunamis are over as soon as one wave strikes. However, one tsunami contains many waves. Sometimes the waves that come after the first hit may be even taller and more deadly. The first waves hit the shore and flow back to sea. They then combine with other waves coming to shore. This can make an even larger wave that washes over land. It may be several hours before all the waves of one tsunami have hit the coast. Sometimes the waves of a tsunami may travel back and forth across the ocean for several days.

Tsunamis can cause a great deal of harm to ocean coasts. Water slowly moves away rock and dirt in a process called **erosion**. Erosion usually takes place over a long period of time. But each tsunami causes a lot of erosion all at once. In one moment, a tsunami can strip a beach of all of its sand. Tsunamis toss huge rocks from the ocean onto the land. They wash away soil and plants on coastlines.

Tsunamis can also kill people. Someone may get washed away in the waves and drown. Flooding from a tsunami may leave people without electricity or clean water to drink.

A wave like this is made when energy travels through water.

ALL ABOUT WAVES

A wave is one way that energy moves from place to place. As a wave moves through the ocean, it does not push the water toward shore. Instead, the wave picks up water particles. A particle is a very small part of something. The energy of the wave causes the water particles to tumble around in circles as it passes through. In a tsunami, these circles reach all the way from the seafloor to the ocean's surface.

Tsunamis are different than other waves. They stretch for thousands of miles across the ocean. They stretch down to the ocean floor, while most waves are close to the ocean's surface.

crest

whitecap

trough

wavelength

Movement of water particles

▲ This illustration shows the different parts of a wave.

Parts of a Wave

All waves have some parts in common. The **crest** is the highest point of a wave. The **trough** is the lowest point of the wave. The distance from one crest to the next crest is the **wavelength,** or how long the wave is. The time between one crest passing a point

Did you know?
Tsunamis were once called tidal waves. Scientists renamed them because the waves have nothing to do with tides.

and the next crest passing the same point is called the wave period.

Tsunamis have long wavelengths and periods. Their wavelength may be up to 200 miles (322 km) with a period up to one hour. An average wind-driven wave has a wavelength of about 500 feet (150 m) and a period of about 10 seconds.

When a wave nears the shore, it slows from the bottom up. The wave falls over, or breaks, when the front of the wave becomes steeper than the back. Breaking waves often have white-foam crests at the top. Tsunamis are unlike other waves. They usually do not break or form white crests. They reach shore as a wall of water or a wave that arches over shore before crashing down on land.

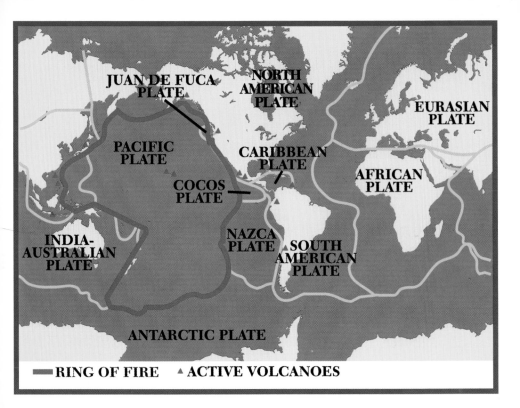

JUAN DE FUCA
PLATE

NORTH
AMERICAN
PLATE

EURASIAN
PLATE

PACIFIC
PLATE

CARIBBEAN
PLATE

AFRICAN
PLATE

COCOS
PLATE

INDIA-
AUSTRALIAN
PLATE

NAZCA
PLATE

SOUTH
AMERICAN
PLATE

ANTARCTIC PLATE

RING OF FIRE ▲ ACTIVE VOLCANOES

This illustration shows different plates and where the Ring of Fire is.

Causes of Tsunamis

Tsunamis start whenever some force pulses energy through a great column of ocean water. The energy moving through the column of water pushes up. It flows outward from the center in large circles, sending waves in all directions. The strength of a tsunami depends on several things. These include the shape of the ocean floor and the depth of the water. It also depends on the amount and movement of energy and the strength of the force that caused it.

Earthquakes are the most common causes of tsunamis. Earth's crust is divided into giant moving pieces called plates. Some plates carry oceanic crust. These plates may crash into each other or slide under each other. The rise or fall of the crust can cause earthquakes under the ocean. This movement disturbs huge columns of water and causes tsunamis. Earthquake tsunamis are often the most dangerous and deadly waves.

Most earthquake-causing tsunamis happen around the Pacific Plate. This plate moves past many other plates. The Pacific Plate's movements have caused many earthquakes. Most of the world's **volcanoes** have formed around its edges. People call the edges of the Pacific Plate the Ring of Fire because of its many volcanoes.

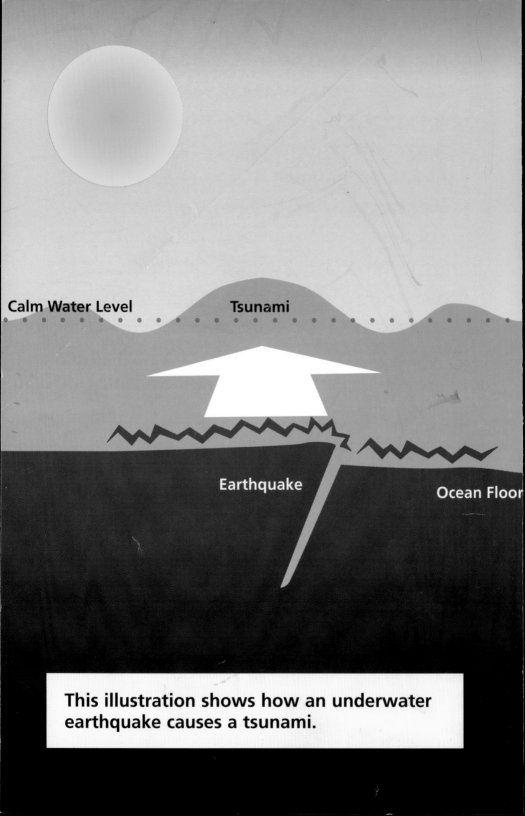

Calm Water Level　　　Tsunami

Earthquake

Ocean Floor

This illustration shows how an underwater earthquake causes a tsunami.

Volcanoes, Landslides, and Meteorites

Volcano eruptions are another main cause of tsunamis. Hot ash, melted rock called lava, and gas flow out of a volcano when it erupts. Volcanoes sometimes cause landslides or huge pieces of volcanic islands to sink back underwater. This causes giant tsunamis. Eruptions can also cause earthquakes that form tsunamis.

The ocean floor has as many different features as dry land. Mountains, valleys, and flat places cover it. Sometimes landslides may fall down underwater mountains and cause tsunamis. Tsunamis caused by landslides often run out of energy before reaching coastlines.

Meteorites falling into the ocean also cause tsunamis. A meteorite is a rock from outer space that has fallen to Earth. This does not happen often.

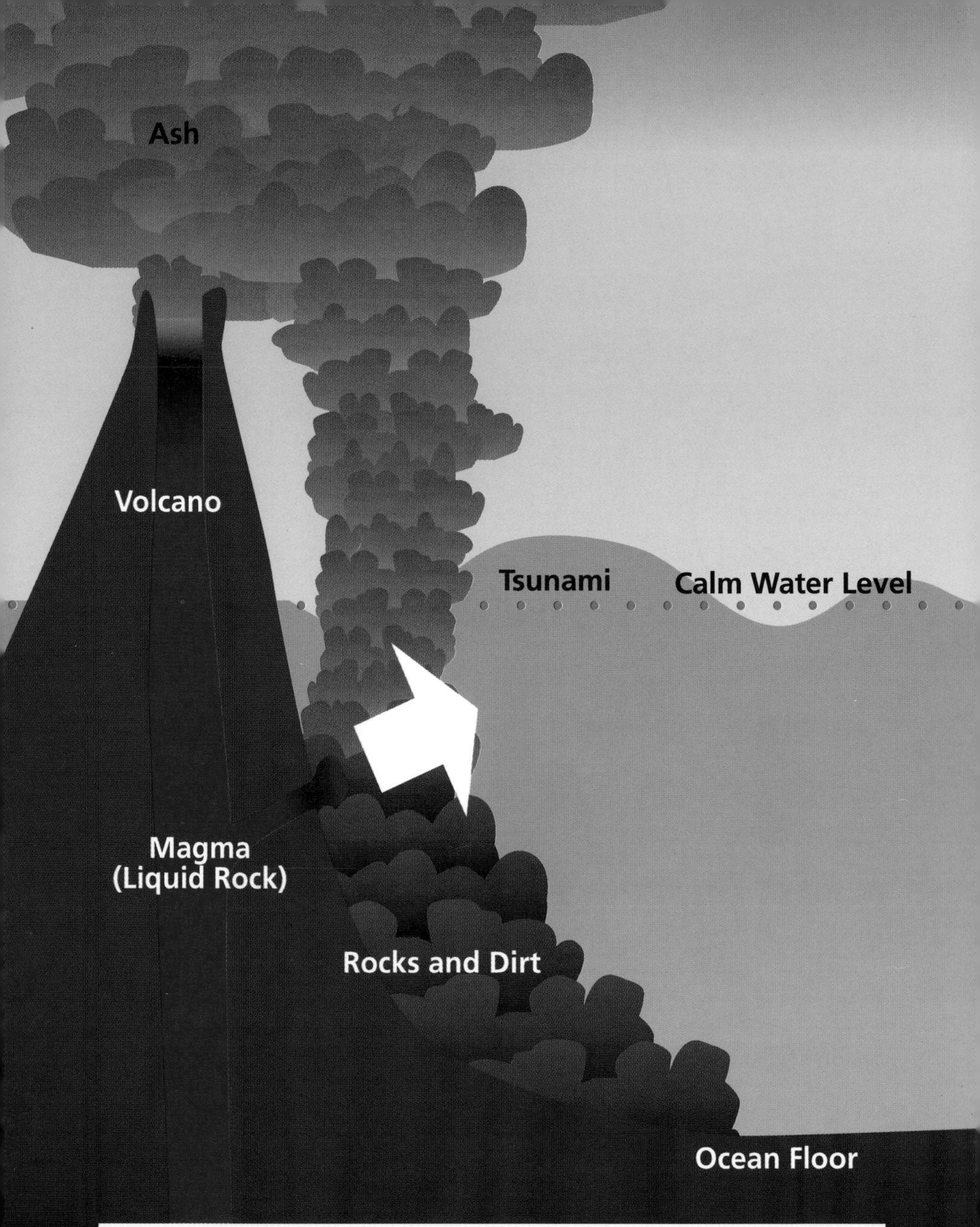

Ash

Volcano

Tsunami   Calm Water Level

Magma
(Liquid Rock)

Rocks and Dirt

Ocean Floor

This illustration shows how a volcano
eruption causes a tsunami.

TSUNAMIS IN HISTORY

Tsunamis have happened throughout Earth's history. People living in the Hawaiian Islands are always in danger from tsunamis. The islands have volcanoes that erupt and cause landslides. The landslides can cause tsunamis.

Hawaiians that lived long ago told this story to explain why tsunamis happen: There was once a magic shark that wanted to sleep forever. He found a quiet place in the ocean and went to sleep. One day, an earthquake woke the shark. He got mad and threw a wave at the nearest island. Every time a volcano or an earthquake wakes the shark, he gets mad and throws a tsunami toward land.

> **This is a picture of the 1946 tsunami hitting shore in Hawaii.**

1946 Aleutian Tsunami

On April 1, 1946, an earthquake shook the Aleutian Islands of Alaska. The earthquake started a huge tsunami that reached land with waves more than 115 feet (35 m) high. In minutes, the tsunami struck shore. It washed boats on shore and smashed houses.

One of the deadliest tsunamis ever recorded struck in 1883. The volcano on the small island of Krakatoa in Indonesia erupted, destroying most of the island. The eruption caused tsunamis that flooded nearby islands, killing about 50,000 people.

It also destroyed a lighthouse, killing all five lighthouse keepers.

The tsunami raced across the Pacific from Alaska. About 4.9 hours later, the tsunami reached Hawaii. People in Hawaii did not know that the tsunami was coming. Some people on shore heard hissing noises as water ran down away from the shore and flowed out to sea. Then a huge wave returned and crashed on shore. The 40-foot (12-m) wave washed over land along the shore. The waves caused $26 million worth in damage.

Altogether, more than 165 people died in Alaska and Hawaii. The loss of life and property made the U.S. government start the Tsunami Warning System. Scientists there try to warn people when a tsunami is coming.

1960 Chilean Tsunami

On May 21 and 22, 1960, two strong earthquakes shook the ocean floor off the coast of Chile. About 15 minutes later, the first 33-foot (10-m) tsunami wave crashed on shore. It washed away buildings and sucked their remains out to sea. As wave after wave hit the shore, more buildings and people were destroyed. About 5,000 people in Chile died from the earthquake and the tsunamis.

These two earthquakes caused tsunamis that spread out and struck coasts all around the Pacific Ocean. Fifteen hours after the earthquakes hit, a 36-foot (11-m) high tsunami flooded Hilo, Hawaii, and killed 61 people. Twenty-two hours later, a tsunami hit Japan. Around 200 people died in Japan when the 20-feet (6-m) waves struck there.

1998 Papua New Guinea Tsunami

On July 17, 1998, an underwater earthquake in the southwestern Pacific near Papua New Guinea caused a large tsunami. Twenty minutes later, three waves washed

The 1998 Papua New Guinea tsunami
flooded this beach, destroying houses.

over part of the northeastern coast.

The waves flooded a 20-mile (32-km)
stretch of beach. The tsunami buried people
under sand and mud. People drowned or
were killed when water smashed their
houses. About 2,202 people died from
the tsunami.

A tsunami destroyed this house, which was built in a low-lying area.

Tsunami Safety

People can do things to stay safe during a tsunami. Scientists from the Pacific Tsunami Warning System call a tsunami watch if an earthquake or landslide takes place in or near the sea. These events might cause a tsunami to strike a certain area. People should stay away from the beach during a tsunami. They should watch and listen to their weather radio or the news for more information.

The Tsunami Warning System gives a warning before a tsunami is about to hit. The warning is broadcast on radio and television and often sirens placed along the coast sound warning blasts. The warning broadcasts tell people when a tsunami will reach their area. People living in low-lying places should move to higher ground right away.

Many towns have tsunami **evacuation zones**. Evacuation zones are areas that people must leave if a tsunami is about to strike. People should leave these zones and go to an evacuation site or a safe place during a warning. They should listen to the radio to make sure all the tsunami's waves have hit before they return home.

Scientists study the shape of coastlines like this one where tsunamis often strike.

SCIENCE AND TSUNAMIS

Today, many scientists study tsunamis. They try to figure out where a tsunami started. This helps them figure out where the waves will travel so they can warn people.

Scientists also study the shape of coastlines and the ocean floor around them. They put this information into a computer and run a special program. The program shows the scientist how tall tsunamis will grow and what part of the coastline the waves will flood.

Based on the computer program, scientists create tsunami evacuation zones in the places a tsunami will most likely flood. Some places, such as Oregon, have laws against putting buildings in tsunami evacuation zones.

Tsunami Warning System

In the Pacific, the Tsunami Warning System is made up of 26 countries. These countries agreed to work together to warn each other of tsunamis in 1965.

The warning system has stations throughout the oceans. A station has instruments called seismometers. These instruments measure the strength and location of earthquakes, which may cause tsunamis.

Some stations are buoys, or floats, placed throughout the ocean. A buoy has science instruments on it. The instruments can measure the waves of a tsunami as they flow through the deep ocean.Machines in the buoys can tell the size of a tsunami by measuring the weight of the water passing over them.

The buoys send this information to a satellite. A satellite is a spacecraft that orbits Earth. The satellite then sends the

This piece of roof is all that remains of a church after it was hit by a tsunami.

information to scientists at the Pacific Tsunami Warning Center.

New tools, such as buoys, help scientists give earlier warnings to people living in coastal areas. By doing this, they hope to stop these killer waves from taking people's lives.

GLOSSARY

crest (KREST)—the highest point of a wave

earthquake (URTH-kwayk)—a sudden shaking of the earth

erosion (i-ROH-zhuhn)—the moving away of something by water and wind

evacuation zone (i-VAK-yoo-ay-tuhn ZONE)— an area people must leave if a tsunami is about to strike

meteorite (MEE-tee-ur-rite)— a rock from outer space that has fallen to earth

reef (REEF)—a strip of sand, rocks, or coral that rises from the ocean floor almost to the surface of the water

trough (TRAWF)—the lowest point of a wave

valley (VAL-ee)—an area of low ground between two hills or mountains

volcano (vol-KAY-noh)—a vent on the surface that allows melted rock, ash, and gas to erupt from inside the earth

wavelength (WAYV-length)—the distance between one crest of a wave and the next

FEMA for Kids: Tsunami
http://www.fema.gov/kids/tsunami.htm

The Living Almanac of Disasters
http://disasterium.com

Tsunami Warning
http://wcatwc.gov/book01.htm

International Tsunami Information Center
P.O. Box 50027
Honolulu, HI 96850

National Water Information Clearinghouse
U.S. Geological Survey
423 National Center
Reston, VA 22092-0001

Pacific Tsunami Museum
P.O. Box 806
Hilo, HI 96721

INDEX